Poplar Grove K-4 Media Center

Cool Collections

Natural Objects

Mir Tamim Ansary

RIGBY
INTERACTIVE
LIBRARY

© 1997 Rigby Education
Published by Rigby Interactive Library,
an imprint of Rigby Education,
division of Reed Elsevier, Inc.
500 Coventry Lane
Crystal Lake, IL 60014

All rights reserved. No part of this publication may be reproduced or transmitted in any form or by any means, electronic or mechanical, including photocopying, recording, taping, or any information storage and retrieval system, without permission in writing from the publisher.

Art director for the series: Rhea Banker

Contributing designers: Susan Darwin Ordahl, Barbara Rusin, Chuck Yuen

Book designer: Susan Darwin Ordahl

The text for this book is set in Garamond Book.

Printed in Hong Kong

00 99 98 97 96
10 9 8 7 6 5 4 3 2 1

Library of Congress Cataloging-in-Publication Data
Ansary, Mir Tamim
    Natural objects / Mir Tamim Ansary.
      p. cm. -- (Cool collections.)
    Includes bibliographical references and index.
    Summary: Offers tips for collecting, organizing, and displaying rocks, shells, leaves, feathers, and other natural objects.
    ISBN 1-57572-115-5
    1. Biological specimens—Collection and preservation—Juvenile literature.
    2. Geological specimens—Collection and preservation- -Juvenile literature.
    3. Biological specimens—Collection and preservation—Exhibitions—Juvenile literature.
    4. Geological specimens—Collection and preservation—Exhibitions—Juvenile literature.
      [1. Biological specimens—Collection and preservation.
    2. Geological specimens—Collection and preservation.]
    I. Title. II. Series: Ansary, Mir Tamim. Cool collections.
    QH61.A57   1997
570'.75—dc21   96-39409
                                                CIP
                                                AC

Acknowledgments
The publisher would like to thank the following for permission to reproduce photographs of their Natural Objects: Rhea Banker, Rory Maxwell, Helen Breen, Jo Ann Hauck, Jennie Nichols, Sangre de Cristo Minerals, Cornelia Tarrant, Alaska and Tucker Burr, Millicent Wakeman, Thomas B. W. Ordahl.

Cover and all interior photographs: Stephen Ogilvy

> **Note to the Reader**
> Some words in this book are printed in **bold** type. This indicates that the word is listed in the glossary on page 24. The glossary gives a brief explanation of words that may be new to you and tells you the page on which each word first appears.

Visit Rigby's Education Station ® on the World Wide Web at http://www.rigby.com

Contents

Collecting Natural Objects 4
Create a Nature Collection 6
Leaves and Flowers 8
Cones and Driftwood 10
Feathers 12
Bones .13
Shells .14
Rocks .16
More Rocks 18
Displaying Natural Objects 20
More Displays 22
Glossary 24
Index .24
More Books to Read 24

Collecting Natural Objects

Do you gather **rocks** and **shells** that **catch** your **eye?** Why?

Maybe **you're** a born **collector.**
Natural objects can be strange, interesting, and beautiful.

But what if you keep bringing things home?
You will end up with rocks in your bed!
Feathers in your drawers! Leaves and dirt!

Create a Nature Collection

Get rid of all that junk! your parents may say. And you might say, *It's not junk. It's a nature collection.*

Prove it! Sort everything into categories. Label them with their scientific names. Set up interesting displays. Make your own nature **museum**.

7

Leaves and Flowers

Autumn leaves are like paintings—colorful! You can frame them or keep them in a photo album. Or try putting a leaf between two sheets of wax paper. Then ask an adult to iron the sheets together.

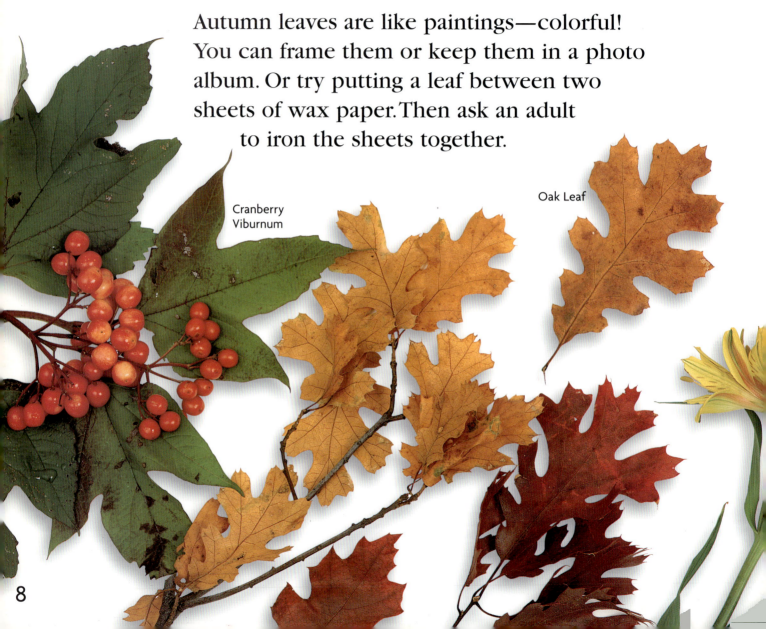

Cranberry Viburnum

Oak Leaf

How many kinds of wild flowers grow in your area? Collect their petals. You may be surprised by how many you find. Petals can be stored in plastic pockets made for photo slides. You can buy these at a camera store.

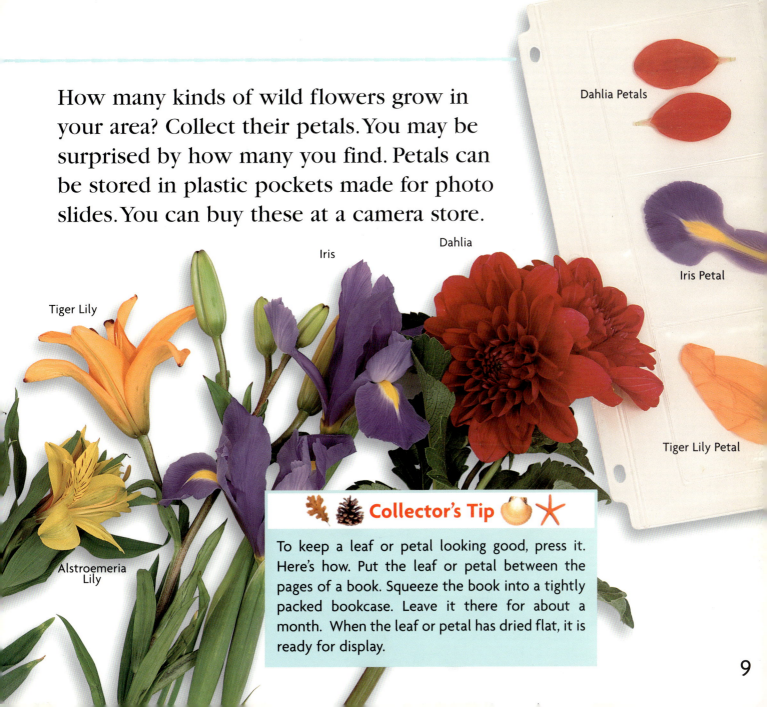

Dahlia Petals

Iris Petal

Tiger Lily Petal

Tiger Lily

Iris

Dahlia

Alstroemeria Lily

Collector's Tip

To keep a leaf or petal looking good, press it. Here's how. Put the leaf or petal between the pages of a book. Squeeze the book into a tightly packed bookcase. Leave it there for about a month. When the leaf or petal has dried flat, it is ready for display.

Cones and Driftwood

Evergreen trees don't change color in the fall. They have needles instead of leaves. Some people like to collect their seed cones. Nature makes the same basic shape, a cone, in so many different ways!

🍂🌰 Collector's Tip 🐚⭐

Take an empty lunch box along on a hike or camping trip. When you see something you want to collect, put it in the lunch box. Plastic snack containers also work well for collecting natural objects.

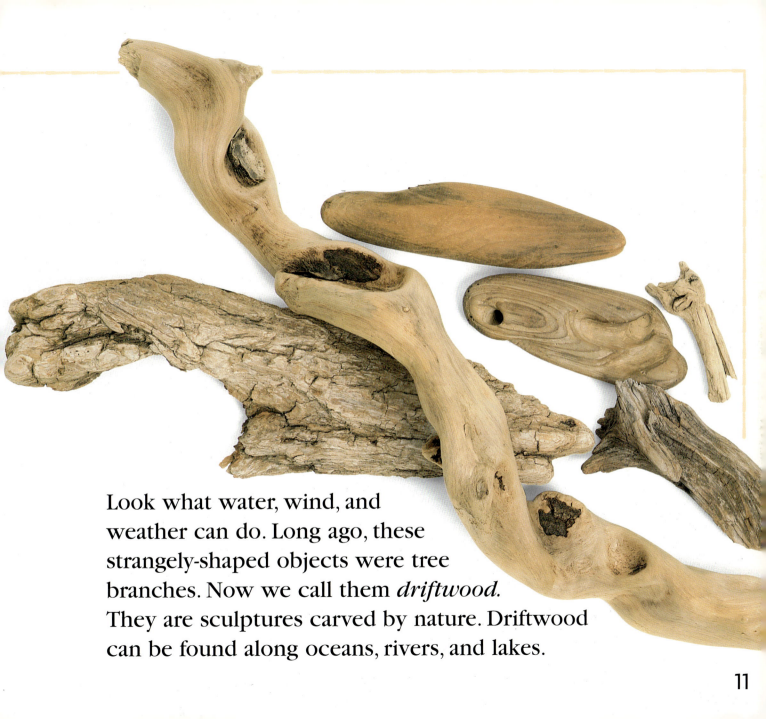

Look what water, wind, and weather can do. Long ago, these strangely-shaped objects were tree branches. Now we call them *driftwood.* They are sculptures carved by nature. Driftwood can be found along oceans, rivers, and lakes.

Feathers

Wherever birds have been, you may find feathers. What birds live in your region? Can you get a feather from each kind? What a great collection that would make!

Hawk

Male Mallard

Female Mallard

Guinea Hen

🌿🌰 Collector's Tip 🐚⭐

A *field guide* is a little book you can take with you on a hike. The book tells what you might see. You can get field guides about plants, birds, rocks, and many other subjects. Before you start your hike, decide what you want to collect. Then go to the library to get a field guide about that topic.

Bones

A bone collector is a bone **detective**. Suppose you find several bones in one place. You will try to fit them together. Maybe you can guess the animal. Would you have guessed what **creature** the skull came from? Did the teeth give you any clue?

Collector's Tip

Never pick up bones unless they are as dry as sticks. The desert is a good place to find such bones. There, the sun bleaches them quickly. Don't touch bones that look fresh. Never pick up dead animals. They may carry germs.

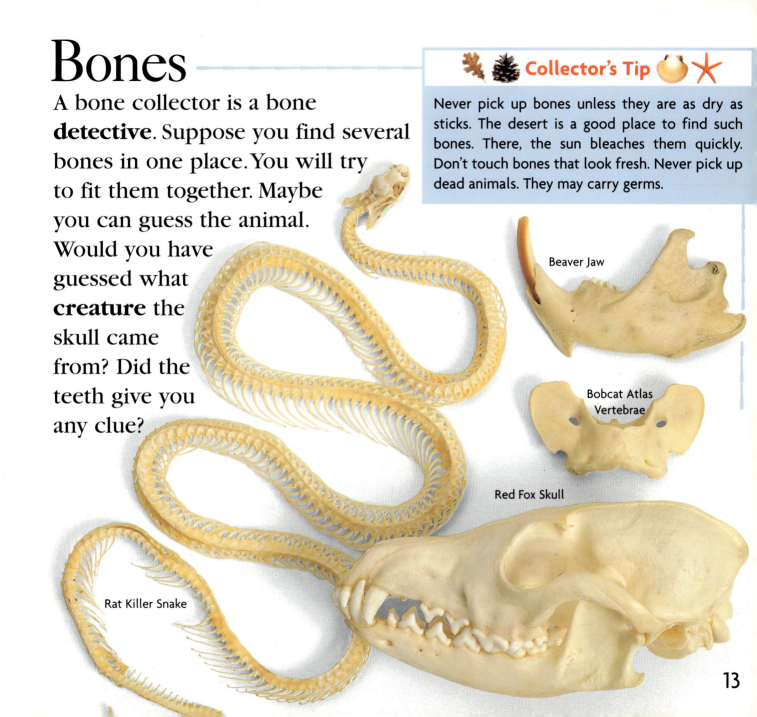

Beaver Jaw

Bobcat Atlas Vertebrae

Red Fox Skull

Rat Killer Snake

Shells

Some creatures have shells instead of bones. Shells come in many shapes. How many different shapes can you collect?

Scallop Shell

Whelk Shell

Olive Shell

Moon Snail Shell

Shells come in many colors, too. That's why people make so many things out of shells—pins, beads, belt buckles, and much more. Shells have even been used as money.

Murex Shell

Turban Shell

Collector's Tip

If you find a shell unlike any other, take it home. Shell collectors call these one-of-a-kind shells **freaks**. Those shells can be quite valuable. Other rare shells include the golden cowry and the glory-of-the-sea.

Rocks

Rocks are easy to find—too easy, in fact. You can't take them all home. How can you choose which ones to collect? Should you focus on rocks with interesting colors? Should you collect rocks with strange shapes?

There are three main types of rock: *igneous*, *metamorphic*, and *sedimentary*. Igneous rocks, such as granite, are made up of crystals and are very hard. Many started out as hot liquids in the earth.

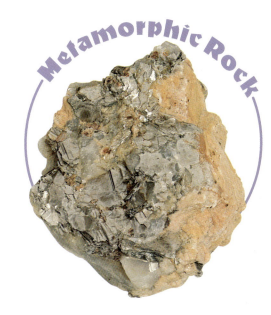
Sedimentary Rock

Igneous Rock

Metamorphic Rock

 **Collector's Tip**

Some rocks are harder than others. To test the hardness of a rock, scrape it with a spoon. If little or nothing scrapes off, the rock is hard.

More Rocks

Sedimentary rocks are grains of something that got pressed together. Sandstone, for example, started out as sand on the ocean floor. The weight of water pressed the sand together.

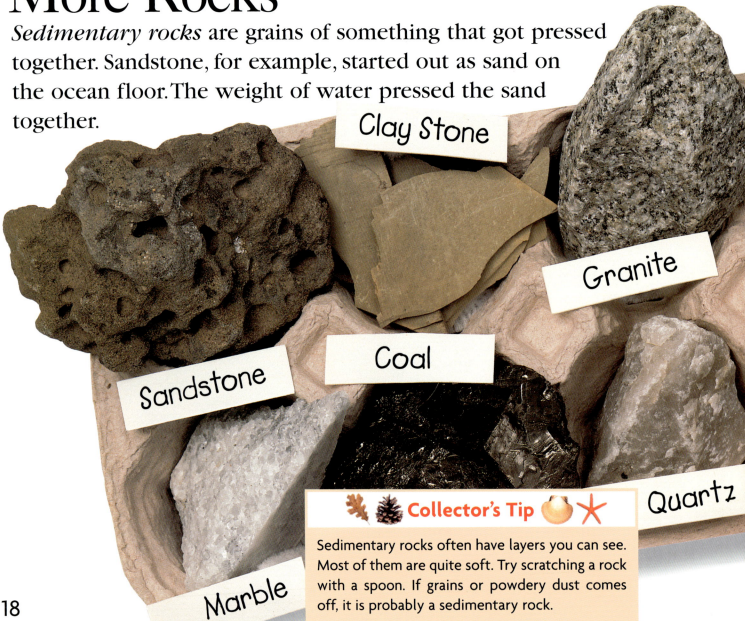

Collector's Tip

Sedimentary rocks often have layers you can see. Most of them are quite soft. Try scratching a rock with a spoon. If grains or powdery dust comes off, it is probably a sedimentary rock.

Scoria

Celestite

Granite

Agate

Galena, Fluorite, Calcite

Iron Staining on Tuff

Collector's Tip

An egg carton makes a good **display** case for small rocks. If you paint the whole carton, it won't look like an egg carton anymore. Choose a color brighter than that of the rocks. That way, the rocks will stand out nicely.

Metamorphic rocks are made of other rocks that were changed by heat and **pressure**. For example, slate was once shale—but it changed. Quartzite was once quartz—but it changed.

Displaying Natural Objects

Here is another way to display natural objects. You can make a collection of items from the same *environment*. An environment is a place in nature and all the things in it. This collection shows you a seaside environment. All the items came from an ocean beach.

This collection gives a picture of a woodlands environment. All the items came from a cold-weather forest.

More Displays

What if you live in the city? Can you still build a collection of natural objects? Yes, you can! Everything in this display came from city yards and parks.

Collector's Tip

You can create a model environment in an empty fish tank. Put some soil in the tank. Then put in moss and other living plants. If you cover the top with clear plastic, you will rarely need to water it. The tank will make its own weather. Such a plant tank is called a *terrarium*.

Shells in a chocolate box display

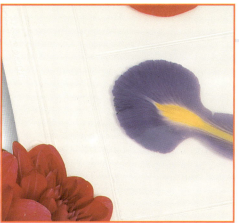
Petals in a photographer's slide sheet display

Woodlands in a jar display

Pine cone on spool display

Rocks in an icetray display

Beach scene in a shoebox display

What happened? Not a thing was thrown out, yet the pile of junk is gone. In its place is this nature museum. Now it's time to invite visitors. Perhaps they can learn something from this collection of natural objects.

Feather in clay display

23

Glossary

Collector Person who collects a certain type of object. Collectors sort, study, and display their collections. 4

Creature Living thing, such as an animal. 13

Detective Person who looks for information that is not easily found. Collectors are detectives because they learn all they can about the objects they collect. 13

Display Used to show off an object in a clear and interesting way. 19

Freaks Rare, or one-of-a kind shells. 15

Museum Place dedicated to the collection, care, study, and display of certain types of objects, such as art, natural objects, dolls, or toys. 7

Pressure Act of a force pushing against another force. Some rocks are formed by the pressure of heat on other rocks. 19

Index

Cones 10-11
Display ideas 7-8, 9, 19, 20-21, 22-23
Evaluating natural objects 12, 14, 16-17, 18
Field guide 12
Igneous rock 17
Metamorphic rock 17, 19
Safety tip 13
Sedimentary rock 17, 18
Sorting natural objects 7-8, 10-11, 12-13, 14-15, 16-17
Tools for collecting natural objects 8-9, 10, 12, 17, 19

More Books to Read

Poskanzer, Susan. *Superduper Collectors*. Mahwah, N.J.: Troll, 1986.
Wong, Herbert H. *The Backyard Detective: A Guide for Beginning Naturalists*. Nature Vision, 1993.

Poplar Grove K-4 Media Center